Pond Life
The Fishing Trip

by Donna Koren Wells
illustrated by Ching

Created by

Distributed by CHILDRENS PRESS®
Chicago, Illinois

Grateful appreciation is expressed to Elizabeth Hammerman, Ed. D., Science Education Specialist, for her services as consultant.

CHILDRENS PRESS HARDCOVER EDITION
ISBN 0-516-08102-0

CHILDRENS PRESS PAPERBACK EDITION
ISBN 0-516-48102-9

Library of Congress Cataloging in Publication Data

Wells, Donna Koren
    Pond life.

    Summary: A trip to the pond provides information about
the plants and animals there and how they interact.
Includes suggested activities such as making a food
chain mobile and collecting snails from a pond.
    1. Pond ecology—Juvenile literature. [1. Pond
ecology.   2. Ecology] I. Ching, ill. II. Title.
QH541.5.P63W45   1990   574.5'26322   90-1644
ISBN 0-89565-581-0

1 2 3 4 5 6 7 8 9 10 11 12 R 99 98 97 96 95 94 93 92 91 90

Pond Life
The Fishing Trip

So come along and find out more about . . .

POND LIFE!

On the first day of summer, Dad woke
Philip up very early. No one else in the
house was awake. Philip and his family
were visiting his grandparents.

"Did you remember that we are going fishing this morning?" asked Dad.
"How could I forget?" said Philip.

7

"We have to be quiet," whispered Dad.
"I'll go and get the poles and bait."
"And I'll get dressed," said Philip.

Philip always liked going fishing at the
pond on his grandparents' farm. He liked
looking for frogs and turtles and ducks.

His father was born on this farm. Philip
began learning about the pond when he
was very little.

"Shhh! Look," said Dad. "There is Grandpa's big old bullfrog. He's the largest frog in the pond now."

"How did he get so big?" asked Philip.
"By eating lots and lots of plants and
insects," said Dad.

12

"He started out as an egg, floating at the
top of the water. Then he hatched into a
tadpole.

"Lots of other tadpoles were eaten by turtles, fish, and snakes. Some were even eaten by insects. But not that one. He changed into a frog. And now he is so large that most other animals leave him alone."

"Look at the turtles," said Philip.
"They like to climb onto the log and
warm themselves in the sun," said Dad.

15

"The sun is important to the life of the
pond. It heats the water all the way to the
bottom and helps the underwater plants
grow."

16

"Do turtles eat plants?" asked Philip.

"These turtles eat just about anything. In fact, they are such good eaters that they help keep the pond clean."

17

Just then, Philip felt a pull on his line.
He yanked his pole. But instead of a fish,
he found a long, green plant.

"I wish there weren't so many plants in
this pond," he said.

"But plants are very important," said
Dad. "They are food for many of the
animals, and they provide hiding places for
little fish and tadpoles.

19

"Be very quiet, and I'll show you a
secret. See? Ducks can hide their nests in
the tall plants along the shore," whispered
Dad.

"When the ducklings hatch, they also use the plants to hide. See how well they match the colors of the plants?"

"What are they eating?" asked Philip.
"That's called duckweed," said Dad.
"And it's a good thing they are eating it.
Because if ducks and other animals didn't,
the duckweed would grow and grow until
it covered the pond.

"Then sunlight would not get to the plants underwater. So you see, all the plants and animals are important to the pond."

"Dad, will the pond still be here when I get to be as big as you?"

"It should be, if we take good care of it," said Dad.

"Dad, I think I've caught a fish!" cried
Philip. "It feels like a big one, too!"

"You *have* caught a big one," said Dad.
"If it were small, we would throw it back
and give it more time to grow and have its
own babies. That's one way we can help
take care of the pond.

"You see, the pond can even feed you.
Let's go show everyone what you've
caught for lunch."

EXPLORE SOME MORE WITH PROFESSOR FACTO!

The sun is very important to a pond. It gives the plants energy to grow. Many little animals eat plants. Other, bigger animals eat the smaller animals. This is called a food chain.

1. The sun gives plants energy to grow.

2. Tadpoles eat plants.

3. Fish eat tadpoles.

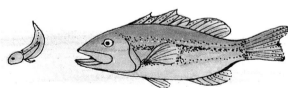

4. People eat fish.

You see, you are part of the food chain too!

You can make a food-chain mobile. Here's how:

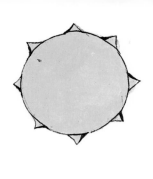

1. First draw a picture of a sun on a paper plate. If you like, you can cut out the rays on the sun, as shown here.

2. Cut out four strips of paper, 2 inches wide and 12 inches long (about 5 centimeters wide, 30 centimeters long).

3. Draw a picture of a pond plant on the first strip, a tadpole on the second, a fish on the third, and a person on the fourth strip.

4. Glue the ends of the plant strip together to make a circle. Attach the circle to the sun, either with glue or as shown. Then loop the tadpole strip through the plant strip and glue it closed. Do the same for the fish strip and the person strip.

5. Ask an adult to punch a hole through the top of the sun. Loop a piece of string through the hole. Now you can hang your food-chain mobile!

Do you know the difference between a pond and a lake? A pond is smaller than a lake. Ponds are not as deep as lakes. The water in a pond is usually very quiet, not wavy. If you were a frog, where would you want to live, in a pond or a lake? Why?

A Snail's Pace

Take a trip to a pond with an adult. Try to collect a few snails. Or, ask a pet store for a few snails. Keep the snails in a jar of pond water with some pond plants in it.

Look at the snails with a magnifying glass. How do they move? Are they fast or slow?

Return the snails to a pond when you are through watching them.

Take a Peek at a Pond

You can make a viewer to help you see underwater in a pond. Here's how:

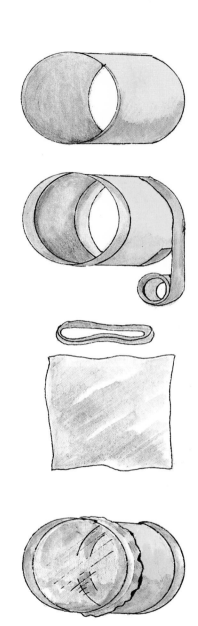

1. First you need a big tin can. Ask an adult to take off the top and bottom with a can opener. Ask the adult to hammer down any sharp points that there may be. Cover the top and bottom edges of the can with masking tape.

2. Cover one end of the can with plastic wrap. Stretch a rubber band around the end of the can over the plastic wrap to keep it in place. Make sure the rubber band is tight around the plastic.

3. Take a trip to a pond with an adult. *Never* go alone. With an adult watching you, push the plastic-covered end of the can a little bit underwater. Look through the other end. What do you see? You will be able to see best on a sunny day.

INDEX